# THE ENCYCLOPEDIA OF KNOWLEDGE:

## WEIRD, CRAZY AND MIND-BLOWING FACTS.

JAMES A. HENDERSON

All rights reserved. No part of this publication may be reproduced, distributed, or transmitted in any form or by any means, including photocopying, recording, or other electronic or mechanical methods, without the prior written permission of the publisher, except in the case of brief quotations embodied in critical reviews and certain other noncommercial uses permitted by copyright law.

Copyright © James A. Henderson, 2024

# INTRODUCTION

Welcome to a world where the banal becomes magnificent and the ordinary spectacular. We explore the amazing fabric of human life in this painstakingly crafted compilation, discovering nuggets of knowledge, whimsy, and surprise. This is a refuge, not just a book. You'll discover echoes of history, innovations, unusual facts, laughs, civilizations, and much more inside these pages.

## What's in store for you in this book?

### 1. *Science Revealed*
Get ready to be amazed by the technological advances, scientific wonders, and innovations that have shaped our world.

### 2. *World Odyssey*
Set off on an adventure across countries and seas. Learn about the peculiarities of other civilizations, the obscure pasts of countries,

and the breathtaking scenery that characterizes our world.

### 3. *Individuals and Exceptions*
Meet the eccentrics, rebels, and visionaries who shaped history in enduring ways. Their tales will entertain and inspire you in equal measure.

### 4. *Historical Enigmas*
Discover the mysteries of bygone eras, conflicts, abandoned towns, and enigmatic relics. Every piece of information serves as a stepping stone toward a larger story.

There is a deeper truth behind trivia: knowledge binds us together. It crosses mentalities, periods, and civilizations. Keep in mind that each truth is a thread in the tapestry of our common history as you turn through these pages. Turn the page, my dear reader, and let curiosity be your guide. Allow your fingertips to stay on the edges; the magic is there.

*Let's*

*go*

*on*

*our*

*adventure!*

A servant of Queen Elizabeth's in the 1600s was so mortified by an inadvertent fart that he fled England for seven years. The Queen met him upon his return, saying, "My lord, I had forgotten the fart."

When his shift came to an end in 2016, a train driver in Spain was unable to recruit a replacement. He legitimately halted the train at a settlement, leaving the 109 people stranded, despite having surpassed his shift hours.

Elephants bury their dead, and decades later, they will return to honor them. They have even been known to bury deceased people.

William Whipple, one of the 56 signers of the Declaration of Independence, thought that one could not fight for freedom and keep someone else in slavery at the same time, therefore he emancipated his slave after signing the document.

To help his kid start seeking work and life, a Chinese father paid gamers in 2013 to murder his son in video games.

When three robbers broke into a woman's home in central Florida in 2010, they believed they had discovered a stockpile of cocaine. They found themselves inhaling the man's and his two enormous Danes' ashes.

A group of more than a hundred enraged ladies marched to the warehouse of a rich merchant in Boston in 1777, demanding the keys and grabbing him by the neck when he refused, all because sugar and coffee were scarce at the time. After unlocking the doors and filling trolleys with coffee, they departed.

An English captain and a Norwegian captain engaged in a 14-hour ship combat in 1714.

After that, the Norwegian commander was running low on ammunition and both ships had sustained significant damage. He requested to borrow some of the English ship's ammunition by sending an ambassador. They declined.

For 3245 years, the seal of King Tutankhamun's tomb remained intact. Up until 1922

In London, Indians possess more real estate than English people do.

A coworker who insisted on giving out book endings was reportedly attacked and hurt by a Russian scientist doing research at Russia's Antarctic station. He was perusing a book.

Compared to a single roll of toilet paper, a bidet consumes a lot less water, energy, and wood, making it a crucial piece of green technology.

Because their brains are working harder at night to fall asleep, people with higher IQs often have trouble sleeping.

When people are in danger, dolphins have been seen to defend them. A gang of dolphins once encircled a surfer from California who was being attacked by a shark and helped him to safety on land. Similar to how they have saved countless lives, dolphins have been mentioned since ancient Greece.

Mike Tyson allegedly made a $10,000 payment to a zoo staff in exchange for the key to a bullying gorilla's cage so he could "smash that silverback's snot box" His proposal was turned down.

You may knot a tie in 177,147 different ways, did you know that? This validates a study conducted by Stockholm University mathematician Mikael Vejdemo-johansson.

It may surprise you to learn that New Zealand is among the few nations devoid of snakes worldwide.

Did you know that Jackie Chan used to cry when he saw a picture of himself?

Situated in Japan's Susami Bay, the world's lowest underwater post box is 10 meters below the surface.

Water was able to flow from the helmets of firemen of firemen's the early 1900s, enabling them to approach the flames.

In the Netherlands, there are more bicycles than people.

There are libraries in Denmark where you may check out a person rather than a book to hear about their life story for half an hour. This wonderful, inventive effort is underway in fifty nations. We refer to it as the human library.

One of the world's longest and steepest stairs is the Jacobs ladder in St. Helena.

A lady was denied entrance into the UK in January 2007 because an immigration officer did not accept that her visit was intended as a week-long vacation in Gateshead.

An American prisoner in Vietnam was dubbed "The incredibly stupid one" by his captors. After coming back to the United States, he gave the names of more than two hundred POWs that he had learned by heart to the tune of "Old MacDonald Had a Farm."

In his will, Ben Franklin gave $2,000 to each of the cities of Boston and Philadelphia to support their youth; but, there was a catch: a large portion of the funds could not be accessed for 100 years, and the remaining funds could not be given out for 200 years. It was valued at $6.5 million in 1990. The funds have supported women's health, firemen, handicapped children, and scholarships.

When Columbus mistakenly believed that manatees were mermaids in 1493, he wrote, "They are not so beautiful as they said to be, for their faces had some masculine traits."
with Denmark, on your birthday, if you're single at age 25, you'll be covered with cinnamon.

In prehistoric Hawaii, women were not permitted to consume certain foods, and men and women ate separate meals. In 1819, King Kamehameha II abolished all religious regulations and carried out a symbolic gesture by dining with women. This is when Luau celebrations originated.

An Australian bartender discovered an ATM bug in 2011 that let him take out far more money than his account balance. He spent almost $1.6 million of the bank's funds over a five-month binge. He paid for his pals' college tuition, hosted extravagant parties, and rented out private aircraft. Afterward, he felt bad and handed himself in to the authorities.

A criminal who has been declared to be "outside of the protection of the law" is not just any criminal; rather, an outlaw is someone who has been condemned to this status and is subject to persecution and death at the hands of anyone.

A distinct breed of horse, Icelandic horses was created in Iceland. Horse importation is forbidden under Icelandic legislation, as is the repatriation of exported animals. One of the strategies used to stop the introduction of new illnesses is the low disease burden that Icelandic horses have in their own country.

When Ben Hsu, a Taiwanese guy, awakened, he had ingested an airpod while he was sleeping. He pressed Apple's tracking button and his body began to bleep. The airpod was in his stomach, as the Kaohsiung Hospital attested to. and then instinctively expelled it from his body. His airpod functioned normally with the battery at 41% after being washed and dried.

While delivering a gun safety class and asserting that "I am the only one in this room professionally enough to handle this block 40," a DEA agent unintentionally shot himself in the thigh.

Abraham Lincoln was a skilled wrestler who won over 300 matches and only lost one before becoming president. He memorably said, "I am the big Buck of this lick," to bystanders. Come on and let your horns ring if you would want to give it a try. The National Wrestling Hall of Fame honors his wrestling accomplishments.

A Cleveland lady was jailed in 2012 after breaking into random homes, cleaning them out, and leaving a bill and her phone number as payment for her services. Breaking in, she cleaned a few coffee cups, removed the trash, vacuumed, and dusted the inside of the home.

A writer, offended by an online review of his book, flew to Scotland, tracked out the reviewer, went to the shop where she worked, and gave her a wine bottle blow to the back of the skull.

A French prisoner broke out from jail using painted nectarines that resembled explosives, all thanks to his wife who had taken flying lessons just for the purpose.

One time, actress Hedy Lamarr? dubbed "the most beautiful woman in the world"?was tricked. In addition, she was a mathematician and the creator of the spread spectrum technology (frequency hopping), which powers Bluetooth and Wi-Fi. She was awarded a star on the Hollywood Walk of Fame in 1960, but the National Inventors Hall of Fame did not induct her until 2014, a full 14 years after her death. Her whole life, she wished that others would recognize her intelligence rather than simply her looks.

The research found that strict parents might make their children more skilled liars because

The eight FIFA World Cup stadiums in Qatar are equipped with massive air conditioners to withstand the intense heat.

The Greeks were appalled when the invading Turks began tearing apart the Partheno's columns to produce bullets against them. They delivered them ammunition with a message that said, "Here are bullets, don't touch the Columns."

Japan's Aomori City is regarded as the world's snowiest city. Over 312 inches of snowfall fall in this city year on average.

With its 250,000 LED lights, the "Interstellar" restaurant in Mexico provides an extraordinary dining experience.

A $20 billion oceanic airport built by Japan is collapsing.

Out of all vegetables, sweet potatoes are considered to be the healthiest.

From a distance, Genting Highlands, a city in Malaysia, seems to be a metropolis floating above the clouds.

The biggest indoor theme park in Abu Dhabi is owned by Ferrari. The Rossa roller coaster, which can reach 150 mph in only 5 seconds, is the fastest in the world.

At 7600 meters, Bear Grylls holds the record for the highest open-air formal dinner party hosted in a hot air balloon.

The Air Lander 10 is an incredible, massive airplane powered by helium. It has individual chambers and can remain in the air for days at a time.

Experts claim that the Blue Javana Banana, which is highly prized in Hawaii and limited to Southeast Asia, has a vanilla-like flavor.

When Morgan Freemark's character in Schawstank Redemption is questioned, "Why do they call you red?" to which he replies, "Maybe because I'm Irish." It's because the book's protagonist was an Irishman with red hair. Freeman was cast for the role, but the phrase was inserted jokingly.

Submarine cables, not satellites, are used to connect to the Internet.

A fox communicates with other foxes by using his height as a signal flag.

With 815 million hectares of forest, Russia is the nation with the greatest forests worldwide.

Taiwan is now the first nation in the world to provide free WiFi to all of its visitors, both domestic and international.

Certain aircraft may continue to fly for up to five hours after losing an engine.

Living in the German village of "Fuggerie" still only costs one euro, and rent hasn't increased since 152.

It will take more than an hour for a stone thrown into the Marina Trench to reach the ocean bottom.

The most costly liquid on earth is deathstalker scorpion venom, which may cost up to $39 million per gallon.

if you were born aboard an aircraft in the United States, even if your parents are foreigners. You are eligible to apply for US citizenship.

One of the world's safest cities is Tokyo, Japan. A six-year-old can use public transportation by themselves.

South Korea is home to the first fully constructed floating metropolis in history.

Faroe Island's Lake Sorvagsvatn juts out over the sea.

Elephant pups learn to feed themselves by placing their trunks into their mothers' mouths.

In the summer, a 10 m deep lake emerges from a dry landscape in Austria. Bridges, benches, and several other nearby trees entirely submerge into the sea.

Rome's Appian Way is a path that dates back to 312 BC. Even though it is more than 2,000 years old, it is still in use. It used to be one of Rome's most significant roads.

With more than 400 skyscrapers now compared to only one in 1991, Dubai is the nation with the highest rate of growth in the world.

In the daring sport of "extreme ironing," participants carry ironing boards to far-off locations to iron garments. In Italy, a man submerged himself 138 feet in the deepest pool in the nation.

By carefully elevating your legs and lying on your back rather than your stomach, you may prevent sinking in fast sand.

The Nokia 1100 is the best-selling mobile phone in history, having sold over 250 million copies.

Japan has an artificial beach. All year long, the inside temperature is completely regulated.

The super bullet train, which was introduced in China, is among the fastest land vehicles. Because it runs on magnets rather than wheels, it is known as the "floating train." It can go at a maximum speed of 620 km/h.

In Japan, specific routes under rail lines are provided for turtles to prevent turtle injuries and train delays.

A small percentage of the people, mostly derived from William the Conqueror, still hold 70% of the land in England.

Cleopatra made a grand entrance as she sailed to meet Mark Anthony for the first time. It is said that she sank her ship's sails in perfume to proclaim her presence before her ship even appeared in the distance. Anthony was astonished and energized by this grand arrival.

A group of students in Scotland visited a contemporary art museum and placed a pineapple in an empty display to test the theory that it was artwork. Not only was the pineapple still there when they came back four days later, but it was also protected by a glass case.

The Katy Freeway in Houston, Texas, is the largest roadway in the world, spanning 26 lanes.

Studies show that sloppy or untidy handwriting indicates great intellect, indicating that your pen cannot keep up with your mental abilities.

Greece's Egremni Beach is among the world's cleanest beaches, with immaculate water.

With 100 carriages and a length of 6,266 feet, the passenger train in Switzerland is among the longest in the world.

In Singapore, there is an Eco bridge that allows animals to cross the road.

Stephen King, whose horror novels rank among the best of all time, must sleep with lights on.

Approximately forty Celtic clans made up Ireland in the 1500s; each claimed territory and often engaged in hostilities. In 1541, certain clans, like the O'Malleys, submitted to English power, while others did not. Being a clan leader was not only risky but also politically complex.

Japan is home to the world's oldest hotel. Run by the same family for 52 generations, the Nishiyama Onsen Keiunkan Hotel in Yamanashi has been in operation since 7:5 CE (1,311 years).

Vietnam debuts the first gold-plated hotel in history. This hotel has gold-colored gates, walls, coffee cups, and even toilet seats.

California is home to both the world's biggest and tallest trees.

Any employee of Philips may work from home that day if he is unable to locate a parking space at the workplace.

Vietnam is home to Hangson Doong, the biggest cave in the world. With its climate, rivers, forests, and clouds within, it is more than 200 meters high.

A group of maritime pirates from the late Bronze Age became the sea people. Their raids primarily targeted Egyptian coastal towns and cities. They are regarded as one of the main factors that led to the end of the Bronze Age, which saw the rise and fall of several ancient civilizations between 1,250 and 1,150 BC.

Rainbows may be seen from an aircraft, and they are complete circles.

In France, there is a road that is only accessible twice a day for a few hours before it submerges under thirteen feet of water.

The ironic thing about Leonardo DiCaprio's relationship history—he starred in movies like Titanic and has dated over eighteen women—is that, despite being over forty, he has never dated a woman who was older than twenty-five.

Milo, who won the boys' division in 540 BC, went on to win six Olympic titles and dominated the sport from the 62nd to the 66th Olympiads. He participated much into his peak, going over 40 at the 67th Olympics, defying age. He also often attended the esteemed Pythian games.

In 480 BC, Queen Artemisia heroically defended the Persians in the Battle of Salamis. To get away from a chase, she established herself as a strategic thinker and led her fleet. She tricked the Greeks into believing she was on their side by smashing a friendly Persian ship. King Xerxes praised her for her bravery.

date of the bottle's expiration Since water never goes bad, it should be used for bottles rather than contents.

The smaller the chili pepper, the hotter it is generally, but color is not a reliable indicator of heat level.

To enable them to descend farther, crocodiles ingest rock.

In the 1950s, women were being issued tickets by police for wearing bikinis on beaches, even though it was against the law in France, Spain, and Italy, among other European nations.

69 is the highest number of children born to a single woman.

An automobile traveling at 60 mph, or 95 km/h, will reach the moon in less than six months.

It has been shown by science that infants learn much more quickly than adults.

The Egyptians would often extract the brains via the nose during the mummification procedure.

If nothing major is done, by 2050 there will be more plastic in the ocean by weight than fish.

Because the F11 fighter plane was quicker than its ammunition in 1953, it shot itself.

Arnold Schwarzenegger had to shorten his workout program for the movie "Canan the Barbarian" because his arm and chest muscles were too large for him to wield a sword correctly.

The world's smallest underground water tunnel Aqueduct Veluwemeer ( Harderwijk, Netherlands) is just 78 ft long.

The national anthem of Spain has no words.

Apple headquarters workers get an average of 125,000 US dollars a year.

The world's first permanent electrified road to charge electric automobiles while driving opens in Sweden.

The Schönbrunn Zoo in Vienna Austria is not only Europe's oldest zoo but also the oldest zoo in the world.

The ancient city of Jericho is the world's oldest wall City with traces of stone fortification reaching back over 9,000 years.

Mount Fuji is a stunning snow-topped mountain and a prominent tourist destination in Japan. It is a volcano that erupted in 1708.

A symposium on women's rights was previously organized in Saudi Arabia, but not a single woman showed up.

In May 2023, 3.2 million saplings were planted by volunteers in the Philippines, shattering the record for the most trees planted concurrently in a single hour.

Broken glass and debris from the Beirut explosion were utilized by a Lebanese artist to build a monument.

The Nilgiri Mountain Train is India's slowest train. The mountainous region between Mettupalayam and Ooty is traversed by the Nilgiri Mountain Train. For this train, a mere 46 kilometers may be traveled in around 5 hours.

The only other Arab nation without a desert is Lebanon.

Having failed to catch on at first, shopping cart inventor Sylvan Goldman was forced to employ models, both male and female, to wheel carts about his store, explain their usage to other customers, and show their value.

In China, there is a city called Tianducheng that was intended to be a duplicate of Paris.

From humble beginnings in Massachusetts, Salem Poor became an American hero in the end. He was born in the late 1740s and paid 27 pounds, or a few thousand dollars today, to purchase his freedom. He quickly enlisted in the American independence movement, demonstrating how quickly one can shift from being a zero to a hero.

To prevent seawater from entering the nation, the Netherlands has constructed walls that span 1,400 kilometers along its coastline. Forty percent of the Netherlands will be underwater if these barriers are taken down.

The Atlantic Ocean is the one that borders the majority of the countries on Earth out of the five oceans. About 133 nations around the world share borders with the Atlantic Ocean.

When grass is cut, it releases a signal to other plants telling them about damage. This signal is the smell of freshly cut grass.

The owner of an American shoe chain decided in 1949 to use his own home as a marketing tool, and ever since he has constructed a shoe house in Pennsylvania that is a highly visible and well-liked location.

Born around 519 BC, Lucius Quintus Cincinnatus was a Roman patrician who worked hard on his small farm until an invasion forced his fellow citizens to look to him for leadership despite his advanced age. In sixteen short days, he won the war, gave up authority, and got back to work on his plow.

Alexandria was founded in Egypt by Alexander the Great in 331 BC. He was recognized as a son of Zeus-Ammon at the Oracle of Siwa. As long as the Egyptians maintained his supply routes for his army, he would not interfere with their customs or impose his ideas of truth and religion.

To prevent themselves from drifting apart, sea otters hold hands while they sleep.

Austrian Hans Steininger is renowned for his beards that break records. He tragically died from a broken neck sustained when he tripped over his beard.

Giant solar-powered lasers have been installed in the Saudi Desert to direct the missing individuals toward water sources.

When patrons keep their phones in a box and refrain from using them while dining, an Australian restaurant offers a 10% discount.

The smallest deer in the world weighs barely 1.8 kg. It looks more like a mouse.

The Akhal-Teke is regarded as the world's most exquisite horse. He appears to have been covered in gold.

Inside the structure, polar bears have made themselves at home on an Arctic island that was abandoned by people a few decades ago.

Reynisfjara is a well-known black sand beach on Iceland's south coast.

To shield residents from noise and air pollution, a five-story apartment building known as "treehouses" is situated in Italy and encircled by 150 trees.

Blue street lights were installed in Scotland and Japan, and these countries have lower rates of suicide and crime.

Because of its concave shape, a London skyscraper was once known to melt cars and set buildings on fire before a permanent sunshade was added.

The world's smallest nation is Vatican City.

There are no mosquitoes in Iceland.

The airport in Brussels is the world's largest retailer of chocolate goods.

Europe's longest coastline is found in Norway.

Lightning strikes can occur up to five times outside of the sun's surface.

To allow vehicles behind to pass safely, Samsung constructed a safety truck in 2015 that shows the road ahead on a screen mounted on the back of the vehicle.

The Mexican walking fish, or axolotl, can regrow missing body parts, including limbs. They can regenerate portions of their brains as well as the same limb up to five times.

The fastest animal on land is the cheetah with a speed of 120 km per hour.

The fastest animal on water is selfish with a speed of 110 km per hour

The fastest animal in air is the peregrine falcon at a speed of 390 km per hour.

Hong Kong is the place where you can find the maximum number of Rolls-Royces per capita in the world.

Trees correctly positioned around buildings may decrease air cooling demands by 30% and can save 20 to 50% in energy consumed for heating.

Mice can sing, but humans can't hear them. Male mice may generate intricate melodies, comparable to songbirds in the ultrasonic region when they spend time with females.

The first rocket emerged in China at about AD 1,330. It features an arrowhead, arrow barrel, arrow feather, and gunpowder tube. The gunpowder tubes were usually fashioned from bamboo

During World War II Dutch minesweepers, and tiny vessels intended to clear or explode naval mines played a key role. Notably, the HNLMS Abraham Crijnssen, a Jan van Amstel-class minesweeper, avoided Japanese soldiers by disguising themselves as a tropical island. Post-war these ships monitor the Netherlands East Indies against revolutions.

The Seawise giant also known as Knock Nevis, was the biggest ship ever constructed in human history; its displacement was 657, 019 tons.

3,000-year-old jugs of honey were uncovered in an ancient Egyptian tomb and it's still tasty.

In Arkansas, blind 9th student Paul Scott participates in cross-country events under the assistance of 4th grader Rebel Hays. Rebel uses a rope to assist Paul and practices tirelessly to keep up with the older rivals.

A US lady discovered her bag that was misplaced by United Airlines 4 years ago while coming home from Chicago. The suitcase reappeared in Honduras.

The high-speed bullet train in China moves so smoothly and vibration-free that it doesn't even reach its maximum speed of

348 km/h if you leave a penny standing inside.

It wasn't until the discovery of America that India had peppers and Italy had tomatoes.

Alexander the Great received reprimands as a kid for using excessive amounts of rare incense at a sacrifice. Years later, he sent his master 18 tons of incense as a gift after capturing Gaza, the primary agricultural supply of that particular incense in the area.

The sunstone is said to be a crystal that the Vikings used for navigation—possibly Iceland spar. They could pinpoint the location of the sun even on foggy days by sensing the polarization of sunlight. They may have arrived in America centuries ahead of other European explorers by using this technique to sail the Atlantic Ocean.

Situated 1.100 feet below Detroit, Michigan lies the Detroit salt mine. It is located 1.500

acres underground and was inaugurated in 1910.

Franz Reichelt, the "flying tailor," was ready to launch his specially built parachute from the Eiffel Tower in 1912, but the leap proved to be lethal.

It was found that James Barry, the renowned British military physician who died in 1865, was Margaret Ann Bulkley. Barry was a trailblazer in his area and had a renowned and highly successful career spanning over 50 years.

Alexander the Great gained his nickname "Great" because of his unmatched military abilities, which included winning every war and personally commanding his troops. He brought Macedonian weapons to India, destroyed the Persian Empire, and laid the groundwork for the Hellenistic world of territorial kingdoms.

It would take you about 21 years to visit every island in the Philippines if you spent one day on each one.

In Sicily, Italy, there is a little hamlet called Nestle that has the strangest human-like form.

Bruce Lee could hurl a single grain of rice into the air and use a chopstick to grab it, complete 50 repetitions of one-arm chin-ups, and perform one-hand push-ups using a storm and index. He didn't belong in this world.

Indonesia has an island called Java. 151 million people are living on this island. It represents around 2% of the global population.

The northern border of the Roman Empire in Britain is marked by Hadrian's Wall, which was built in 122 AD under the command of Roman emperor Hadrian. This

73-mile wall was built as a show of Roman might as well as a military structure in Northern England. It served as a barrier against invasions and a way to control commerce and travel with its forts, watchtowers, and gates.

Two independent businessmen in Switzerland established rival milk firms in 1866. After both men passed away in 1905, the two businesses combined to become Nestle.

The tribe got power when Kenyan Faith Kipyegon won the gold medal in the women's 1,500-meter race in the Rio Olympics.

More than 798 kg of winnings are needed for Formula 1 racing vehicles. Additionally, the driver's weight should not be less than 80 kg. You must place a weighted item on the driver's seat to raise it to the recommended weight if the driver weighs less than 80 kg.

A youngster was allowed access to the set of Thor: Ragnarok by the Make-a-Wish Foundation. The famous phrase, "He's a friend from work," appears in the scene when Thor enters a gladiatorial arena and first encounters the Hulk. The child genuinely proposed this language.

Uruguay's circular bridge was constructed to allow cars to take in the scenery at a slower pace.

When agreeing to shoot a film, Steve McQueen often demands free stuff excess from studios, such as pants, electric razors, and other goods. Later on, it was found that he had given these items to the Boys Republic Reformatory School, which he had attended as a teenager.

A rail station in Japan had planned to shut down but found a student was the only one utilizing it. After her high school graduation, they decided to part ways.

The world's most mountainous nation is Bhutan. Mountains encompass 98.8% of the land.

If you stayed in every hotel room at Disney World for a single night. You'll need 68 years to sleep through them all.

The shoreline of Lake Constance extends into Germany, Switzerland, and Austria.

Submerged for 1,200 years, the Lost Egyptian city of Heracleion was discovered.

A 19-year-old student created an ocean purification system that can eliminate 7,250,000 tons of trash from the global ocean.

Bermuda has to gather rainfall to thrive since it lacks freshwater springs, rivers, or lakes. White steps regulations were imposed on homes more than 400 years ago to collect rainwater and channel it into subterranean reservoirs. Every house can support itself. Both main water and water rates are absent.

Chile is home to the biggest swimming pool in the world. With a length of 1,013 meters (3, 323 feet), it holds about 250 million gallons of seawater.

The world's first practical helicopter, the VS-300, made its flight in Stratford, Connecticut, on September 14, 1939. The helicopter was the first to use a single main rotor and tail rotor configuration, having been created by Igor Sikorsky and manufactured by the US aerospace corporation's Vought-Sikorsky aircraft subsidiary.

Research on food and gender found statistically significant differences. Whereas women favored healthy grains and vegetarian chocolate, males were overwhelmingly drawn to meat. Men tend to be hungry before supper, but women tend to be hungry first thing in the morning and are more prone to nibble all day.

Renowned scientist and Nobel laureate Richard Feynman was also a proficient bongo player. He would often play the bongos to decompress and find inspiration in the drum rhythm and reading. It's similar to having a scientist who enjoys listening to cosmic music.

The pioneering work on color theory by Scottish physicist James Clerk Maxwell was motivated by his vivid dreams about colors. He created the first color photograph and the theory of color perception, which laid the groundwork for contemporary color science. It is comparable to having an artist use one's mental palette to paint.

Throughout their lifetime, 12 bees will need to harvest one teaspoon of honey.

Nikola Tesla was terrified of germs and was completely preoccupied with hygiene. It was like having a genius who was afraid of germs in the lab because he wouldn't touch jewelry, hair, or anything else he believed might be contaminated.

The eccentric scientist and electrical engineer Nikola Tesla had a great affection for pigeons. He once said that he had fallen in love with a pigeon and that pigeons were messengers from the universe who would provide him with thoughts and inspiration from space.

Despite having six legs, dragonflies are unable to move for extended periods due to the weakness of their legs.

The Czech term "robot," which means "forced labor" or "drudgery," is where the word "robot" originates. Karl Çapek, a Czech dramatist, initially presented it in his 1920 play "R.U.R. ( Rossum's Universal Robots)."

It was not for his renowned work on relativity that Einstein received the Nobel Prize in physics in 1921, but rather for his discovery of the photoelectric effect. The basis of quantum mechanics is laid by the photoelectric effect, which explains how a substance emits electrons when exposed to light.

Australian native Myrmecia Midas, or nighttime ants, are said to utilize polarized moonlight to locate their route, according to Macquarie University in Sydney. After dusk, these ants left their nests to search in trees for food, returning home in the gloom of morning.

Distraction might be a contributing factor in overindulging in food. Research indicates that persons who are distracted during a hedonic activity are likely to enjoy food less than they would if they were focused. This might cause people to feel unsatisfied and increase their intake to make up for the difference in pleasure.

The design of the tablet was conceived in the mid-1900s (Stanley Kubrick illustrated imaginary tablets in the science fiction film 2001: A Space Odyssey, released in 1968), and a prototype was created in the last two decades of the same century. The first mass-market tablet to attain broad appeal was the iPad, which was launched by Apple in 2010.

More electrical impulses are produced in a single day by one human brain than by all of the world's telephones put together.

With more than 1,000 patents to his name, Thomas Edison is the inventor of the phonograph, motion picture camera, and electric light bulb.

According to research, when two people are really in love, they synchronize their breathing and heart rates for three minutes while staring into each other's eyes.

A condition known as "Stoneman syndrome" causes injured tissue to turn into bones. They are the reason for the victim becoming more and more like a statue over a long period.

A USB condom is a tiny gadget that only allows power to flow through it, preventing data transmission when connected to a smartphone or tablet with charging capabilities in public areas.

Male ostriches in ostrich farms often find their human caregivers more beautiful than female ostriches, which makes it harder for them to reproduce.

Natasha Demkina, a Russian lady who went by the nickname "the girl with x-ray eyes," rose to fame across the world for her assertion that she could see into the human body without the use of medical technology. Natasha was born in Saransk, Russia, on January 11, 1987. At the age of ten, she realized she had a special talent.

The memory capacity of an adult human brain is 2.5 million GB on average. It doesn't run out of storage space, however. There are many distinct types of memories stored in a single human brain, and the amount of memories humans can physically store is infinite.

Bacteria use a process called quorum sensing to coordinate their actions and behaviors by exchanging chemical signals with one another.

Philo Farnsworth, the man who invented television, forbade his kids from watching because he thought it would make them dumb.

If their parents are absent or unable to care for them, squirrels will adopt the young of other squirrels, often from the same social group.

Studies have shown that ginger may be good for sexual health. Research indicates that increasing reproductive hormone levels, decreasing oxidative stress, and improving blood flow may all help to enhance sexual desire and performance.

A human might fit through the arteries of a blue whale's massive heart.

Underwater, humpback whales may be heard singing intricate, 30-minute melodies across great distances. These songs are used for mating, communication, and navigation, among other things. Every population has a distinctive tune that changes with time.

Cryovolcanoes are amazing characteristics of cold worlds in our solar system. They are discovered in space and are generated when chemicals like water, methane, or ammonia freeze and erupt as frozen vapor forming volcanic snow.

Even though oxygen is necessary for life, breathing 100% oxygen at atmospheric pressure for prolonged periods may be harmful and result in oxygen poisoning syndrome.

There are homes perched above shopping mall roofs in China.

Penitentes are snow-covered structures that may be seen in Chile and other high-altitude areas. They are made of tall, thin spires of ice or snow that have been eroded and sublimated. Certain climatic circumstances, such as desert or high-altitude regions, give them their distinct form.

As far as we know, the deepest place in the ocean is the Mariana Trench. This trench descends around 11 kilometers (or 7 miles) into the western Pacific Ocean.

The Braille method was created in 1829 by the blind French professor Louis Braille. By enabling people with visual impairments to read and write on their own, this tactile writing method transformed reading and communication. Raised dots are used to represent letters and numerals.

Magnificent atmospheric occurrences known as "light pillars" may naturally transpire in frigid regions. Light pillars are rotating vertical columns of light that reach into the sky as a result of interactions between artificial and natural light sources and suspended ice crystals.

The lack of gravity in a microgravity setting, such as space, can certainly impact physiological processes, including digestion. Burping is less prevalent when gas does not ascend to separate from liquids and solids in the stomach due to the absence of gravity's pull. Alternatively, a mixture of gas bubbles and liquid may cause pain. The absence of gravitational directionality makes vomiting in space more difficult.

A dog in Colombia started paying for cookies at a shop with a leaf in his mouth after seeing consumers trading money for products. The charming demeanor of the shop personnel encouraged them to comply, and they now take the present as money, enabling the dog to purchase cookies daily.

Lenovo CEO Yang Yuanging received a $3 million incentive in 2012 in recognition of the company's very strong financial results. Rather than adding it to his, he divided it among 10,000 low-level workers, such as assistants, receptionists, and manufacturing line workers, who took home an average of $314 apiece. In 2013, he repeated the same action.

Elmer Alvarez, a homeless man, discovered a $10,000 check written out to Roberta Hoskie, a real estate broker, in 2018. Alvarez located Hoskie and gave her the check back.

Hoskie was moved by this act of kindness and helped the man get housing, cover his rent, and pay for his education. Alvarez was appointed to the board of directors of her firm after they started working together.

To prosecute a chemical firm for contaminating his town, a Chinese farmer who left school in the third grade spent 16 years educating himself on the law. Wang Enlin, who couldn't afford the local bookstore's legal books. paid the owner sacks of maize in exchange for the right to sit, read, and painstakingly copy the data while using a dictionary to help him comprehend. In 2017, his perseverance paid off as he prevailed in the case.

Dale Schroeder, the Lowa Carpenter, led a thrifty lifestyle, owning just two pairs of pants, and worked for the same firm for 67 years. He never got married or had kids, but he did accumulate $3 million, which he bequeathed to support 33 students' college tuition when he passed away.

A wooden stirrer that was unintentionally left overnight on his doorstep in 1905 by 11-year-old Frank Epperson was filled with soda powder and water. Frank discovered his drink iced up like an icicle when he stepped outdoors the next Monday after a chilly night. He began distributing the dessert throughout his neighborhood since he thought it was good. After producing the frozen dessert for 20 years, you eventually received a patent for "popsicles."

A software flaw in video poker machines was discovered in 2009 by John Kane, a US citizen, who took advantage of it to pay out winning hands at a significantly greater stake amount than he had placed. He was free to retain the money since all you were doing was pushing buttons that he was permitted to push.

Crows are capable of identifying human faces, and I've heard that they harbor resentment against those they find objectionable.

At a music festival in 1979, the Italian punk band Skiantos brought a kitchen, a table, a TV, and a fridge. They cooked some spaghetti and ate it without making any music.

When the child was only two days old, she was abducted from South Africa. She was reunited with her biological family 17 years later when she happened to enroll in the same school as her younger sister. The girls were quite similar to one another and became friends very immediately. Their connection was proved by a DNA test.

The pilot passed out in 2001 while operating a small hired aircraft with his family. After that, Rowan Atkinson, also known as Mr.

Bean, took over piloting the airplane until the pilot came to a few minutes later and made a safe landing.

Finland has used a creative approach to almost end homelessness nationwide. The Method entails providing them with a modest residence and unconditional mental health assistance. Consequently, eighty percent of program participants go on to successfully reintegrate into society, finding work, secure housing, sobriety, and peace of mind.

After serving 23 years in the US Marine Corps, actor Rob Riggle retired as a lieutenant colonel. Among other decorations, he received a combat action ribbon during his deployments in Kosovo, Liberia, and Afghanistan.

Because of his great affection for animals, Leonardo da Vinci would often buy case animals and give them away.

To summon her offspring to devour her alive, a mother black lace-weaver spider will beat out rhythms on her web.

A dead mosquito was discovered by Finnish police while they were pursuing a stolen automobile. To identify the thief, they utilized blood from the mosquitoes' most recent meal that had been examined.

In 1760, an astronomer went to India to see Venusian transits. He missed the date because unfavorable winds drove his ship off track. He spent eight years waiting for the next Transit of Venus in India, but on the day of the event, the sky clouded up and he was unable to see anything. after he got back home. He discovered that he was being replaced in his position and had been deemed legally deceased. All of his family pillaged his estate, and his wife remarried.

The only location where we may see complete sun eclipses is on Earth. This occurs because, while seeming the same size to us, the sun and moon are 400 times larger and 400 times further apart.

In 1988, the authorities decided that a large portion of the history taught to over 53 million Soviet schoolchildren was false, leading to the cancellation of their final history examinations.

The sea slug species Cyerce nigricans is also referred to as the "black dragon nudibranch" or the "black sea slug." Its strikingly black coloring and transparent body define it. This fascinating creature lives in tropical seas in the Indo-Pacific area, where it consumes a variety of algae types.

The goal of the 1929 experimental train system known as the George Benny transportation system was to transform transportation. Create with less traffic and more efficiency in mind. It suggested building an elevated train network to facilitate mobility. It added to the continuous investigation of creative transportation solutions in the early 20th century, even if they were never completely implemented.

The echinoderm family includes starfish, which are often confused for fish. They depend on a dispersed neural system instead of a centralized brain. They use a circulatory system based on water circulation instead of blood. Despite these distinctions, their special adaptations allow them to flourish in a varied maritime habitat, demonstrating nature's amazing capacity for creativity and ecological niche-fitting.

The difference engine and the analytical engine, two of Charles Babbage's ground-breaking discoveries, introduced the idea of mechanical computing. These devices set the stage for the development of calculators, the understanding of logarithms, and eventually the creation of contemporary computers, which fueled the evolution of technology.

The first printing press in European history to be developed and constructed by Johannes Gutenberg. One of the first books ever printed with moveable type was the Gutenberg Bible, which he produced in 1455.

The uncommon Madagascar primate known as the Aye-Aye looks like a mix between a rat and a raccoon. Its striking characteristics include very long, bony fingers, a bushy tail, and huge eyes. These fingers are used to tap on trees to find and retrieve insects, displaying their distinct way of foraging in the wild.

Since earthworms have both male and female reproductive organs, they are hermaphrodites. During mating, they may procreate by trading sperm with another earthworm. Their capacity to multiply effectively and support healthy soil and nutrient cycling is guaranteed by their special reproductive technique.

The only animals that can fly continuously are bats. Their ability to fly and maneuver in the air with remarkable agility is attributed to the thin membrane that covers their elongated finger bones.

Because they can taste the air, cats often curve their lips when they use their tongues to sample the millions of pheromones that humans are not able to detect. The Flehmen answer is the name given to this curl.

In the 19th century, firefighters' masks resembled Star Wars figures.

Native to Indonesia, Papua New Guinea, and the Philippines, rainbow eucalyptus trees display an amazing display of colors. Their bark flakes off in stripes, exposing vivid shades of orange, red, purple, blue, and green that captivate the eye and enhance the attractiveness of a tropical setting.

The oxidized iron in briny salt water gives the bloodfalls in Antarctica their characteristic crimson hue. This unusual occurrence happens when water that is rich in iron emerges from under the Taylor Glacier, producing a breathtaking natural display in the freezing environment.

At Bell Laboratories, John Bardeen, Walter Brattain, and William Shockley created the point-contact transistor, the first functional transistor, in 1947. The first widespread use of transistors occurred in the early 1950s with Shockley's 1948 introduction of enhanced bipolar junction transistors, which went into production.

When it comes to hunting, powerful male lions are in charge of defending the pride's territory, with female lions accounting for 90% of kills on average. The way the female lion behaves goes against the gender norms that are often seen in other animal species where males dominate while hunting.

Diamonds and snails are both often used in jewelry. The Cuban snail was almost exterminated via hunting for industrial uses, but it is now protected on the official list of endangered species.

In a sizable burial mound at Oseberg Farm in Norway, a wonderfully preserved Viking ship from the early ninth century has been found.

The brains of dolphins and humans are highly developed, with intricate architecture and distinct areas. Dolphins' brains contain specific regions for auditory processing and echolocation, which are essential for underwater communication and navigation, whereas human brains have bigger, more folding neo-cortices for higher cognitive functions.

Our eyes move quickly in different directions during the rapid eye movement (REM) stage of sleep, which corresponds to the visual content of our dreams. For example, our eyes may follow our dream motions if we dream of sprinting.

A house fly utilizes its tarsi to taste and access the surface's chemical makeup when it settles there. The fly is more likely to feed on a surface if its foot sensors pick up sugar or other alluring substances.

Insects have a network of microscopic tubes called trachea that carry oxygen directly to their tissues in place of lungs. Spiracles, tiny apertures along the side of the insect's body, allow these tracheal tubes to open out to the outside world. Every cell in the insect's body receives oxygen once it reaches the spiracles and diffuses along the trachea.

Around the age of 78, Benjamin Franklin devised bifocals. Historiographic stories claim that he became weary of having to alternate between two pairs of glasses—one for reading and the other for distant vision—all the time. Franklin split lenses in half and merged them to make a single pair of spectacles that gave him near- and far-sighted vision to get around this problem.

China has enormous man-made caverns known as the Yungang Grottoes, which were constructed in 200 BC. The precise strategies used in its creation are still unknown since there are few historical documents describing the processes of building.

Umbrellas were not utilized for rain protection until much later, in the 16th century in Europe. Originally seen as a luxury item, ladies mostly utilized them to protect themselves from the rain while traveling.

After all, kangaroos can't go back. because their rear legs have specific joints and muscles that are geared for forward mobility, allowing them to propel themselves forward with great strength. Their long tails also act as a counterbalance as they hop, which helps them go forward even more but helps them travel backward less effectively.

In order to shield themselves from the sun's harmful UV rays when they eat on leaves high in trees, giraffes have developed long, black tongues. It is believed that the dark coloring is caused by melanin, which acts as a natural sunscreen. They can also grab and move items with their tongues, which makes them quite helpful for detaching leaves from trees.

Interesting amphibians, and mudskippers may be found on tidal mudflats and mangrove swamps across the Indo-Pacific area, which includes parts of the Caribbean. With the help of their powerful bodies and pectoral fins, they can climb trees. This facilitates their ability to find new eating sites, avoid predators, and control their body temperature.

Because of their very keen sense of smell, polar bears can locate seals more than 30 kilometers away, even when they are hidden by thick ice.

According to a NASA satellite image, a makeshift lake in Death Valley's poor water basin is demonstrating its durability. Following Hurricane Hilary's formation in August 2023, the lake steadily decreased in size but continued to exist throughout the autumn and winter. In February 2024, a strong atmospheric river filled it back up.

There is no blue pigment in the skin of a blueberry. The skin's heartburn nanoscale structure, which scatters blue light more than other wavelengths, is responsible for the color blue.

Hummingbirds walk with little efficiency. Because of their very small leg bones, they don't have much support while walking. Moreover, their center of gravity has been moved to their chest, which makes walking uncomfortable for them.

While they go by "raft" while on the water, pigeons are referred to as "waddle" online.

Your brain is the fattest organ in your body since it contains at least 60% fat.

The dandy horse, often referred to as the hobby horse or draisine, goes back almost a century before the 1930s. Carl Drais came up with it in 1817. This ancient bicycle was a significant forerunner to the modern bicycle as it lacked pedals and needed users to propel themselves forward with their feet.

Legend has it that when Queen Margherita visited Naples in 1889, a local pizzaiolo by the name of Raffaele Esposito made a pizza with toppings that symbolized the three colors of the Italian flag: green (basil), white (mozzarella cheese), and red (tomatoes). The pizza was later dubbed the Margherita pizza, and it is now regarded as one of the most recognizable and popular varieties in the world.

Australian rower Bobby Pearce, who went on to win the race in 1928, paused in the middle of the race to let a family of ducks pass in front of his boat.

Drivers in Germany are required by law to move to the sides of the road when traffic stops to provide space for emergency vehicles.

Bamboo bottles are a fantastic and biodegradable substitute for plastic bottles. India's Sikkim state has already made bamboo water bottles available and outright outlawed the use of plastic bottles.

In 1926, Nikola Tesla made a prediction that would eventually become a contemporary mobile phone: "We will be able to communicate with one another instantly, regardless of distance." Furthermore, we will be able to see and hear each other just as well as if we were in person. A guy will have enough room in his pocket to hold one."

Detroit Electric's electric vehicles were able to consistently go 80 miles between charges more than a century ago. An automobile covered 211.3 kilometers on a single charge in one test.

Of all the automobiles Rolls-Royce has ever made, 75% of them are still in use today.

The Gulf of Alaska has a location where two oceans converge but do not mingle.

Because Mars' atmosphere is so thin, your head would be at 0°C and your feet would be at 24°C if you were to stand on the planet's equator at midday.

There is less gravity in certain portions of Canada, such as the Hudson Bay Area and sections of Quebec, thus if you go there, you will weigh less than anywhere else in the globe.

When a space shuttle launches, the majority of the smoke that rises is not exhaust. To absorb the acoustic shock waves that may otherwise destroy the shuttle, water vapor from the pool of water below it is used.

The need for plastic bottles is eliminated with a water bubble known as "Ooho" that dissolves in your mouth. This is the drinking water of the future.

Photographer Image When Chan happened to snap a picture of a crow riding on the back of a bald eagle, it was an opportunity of a lifetime.

When unsure if the environment is secure and devoid of predators, penguins have been known to force one another into the ocean.

President Andrew Jackson paid off the whole national debt on January 1st, 1835, the first occasion in American history to do so.

There's this restaurant in New York that uses Grandma's instead of chefs. Every day, a new grandmother from somewhere in the globe creates her meal.

Iran is home to a pencil store. The owner understands where to look among the thousands of pencils to pick the right one.

A 4-kilometer railroad track called "The Tunnel of Love" winds through a thick forest in Ukraine.

A sport known as "banzai skydiving" exists. First, use the parachute to exit the aircraft, and then make the subsequent leap.

Fifteen cardiologists arrived to rescue a lady who had a heart attack while traveling from Manchester to Florida after the flight attendant requested assistance. They were taking a plane to a seminar on cardiology.

The only creatures that never drink water in their lives are kangaroo rats.

A two-foot cycle in Portland is the world's smallest park, according to the 1971 Guinness Book of Records. Mill End Park is the name of it.

Dubai is home to the first underwater tennis arena in the world.

Just 14% of the 3,150 tonnes of gold produced annually worldwide come from China's 440 metric tons of annual production. With 245 metric tons, the USA ranks fourth in terms of production.

A female With a math PhD, he was able to decipher the ticket creation process and win the fourth scratch-off jackpot.

The world's biggest desert is found in Antarctica. The Antarctic continent is covered by the 5.5 million square mile Antarctic Polar Desert.

An uncommon apple with a jet-black color is the black diamond. It is unique to Tibet and has a sweeter flavor than honey.

Initially, clownfish are male. A female clownfish coexists with a bunch of men, and upon her death, she mates with only one of them to become the next female.

The mango is one of the fruits with the greatest sugar content. About 45 g of sugar may be found in one ripe mango.

High school in Japan where pupils study for university examinations (Gaokao) from 6:20 a.m. to 11:00 p.m.

The Australian millionaire Cliff Palmer has declared his intention to launch a perfect duplicate of the Titanic by 2027.

Approximately 20 million tons of gold, or over 700 trillion dollars, are contained in the ocean.

An early predecessor of the elephant with a large mouth and tusk resembling a shovel is the platybelodon.

Because female flight attendants are lighter than males, the aviation industry may save up to $500k in fuel annually by hiring them.

The bee visits at least 1,000 flowers in its less than 40-day life, and it yields less than a teaspoon of honey.

In Terminator 2, Arnold Schwarzenegger received almost $21,429 for each syllable he said.

The world's most visited cities are Singapore, Bangkok, Paris, London, and Dubai.

There's a cinema theater in Switzerland where you can watch movies on twin beds instead of seats.

In Barrow, Alaska, the sun sets on November 18 and rises the next year on January 23. There are 67 days of complete darkness throughout this time.

India was one of the wealthiest nations on earth before the 18th century. We called it The Golden Bird.

With 221,800 islands, Sweden has more islands than any other nation in the world.

In the summer, the Eiffel Tower may rise 15 centimeters higher because the iron framework expands as it becomes hotter.

The eye is the second most complex organ in the human body after the brain, with estimates indicating that it has around 2 million functional pieces.

Compared to wealthier individuals, poor folks are more giving. According to polls, those with lesser incomes donate a greater percentage of their earnings to charitable causes.

The majority of chuckles you hear on TV these days were captured in the 1950s. It's official—you're listening to the laughing of the dead.

More people travel on Indian railroads than live in Australia as a whole.

Hachiko was a devoted dog that, despite his master's passing, waited for him to return for more than nine years outside Shibuya Station.

In a single second, a large bird flaps its wings 230 times. This indicates that a bee's wing muscles are equal to an automobile's engine.

Switzerland, often referred to as the Earth's paradise, is the most breathtaking place in all of Europe.

Uranus can accommodate 63 earths.
A tree that has leaves that are darker than the surrounding foliage may be hiding a corpse.

In 1891, Thomas Edison's 1876 invention of an electric pen was modified to create the first tattoo machine.

The "Genetic disc" is an amazing relic that has the power to change the course of history and the very beginnings of humanity. With the use of a microscope, one may see the amazing information contained on the site-carved disk. The piece, which was discovered in South America, is thought to be older than 6,000 years since it does not fit into any known ancient American civilization.

The most birthdays occur in August, the fewest in February, and the majority of serial murderers are born in November.

Your brain uses a technique known as unconscious selective attention to choose to ignore your nose, even though you can always see it.

The Statue of Liberty's seven spikes on its crown stands in for the world's seven continents and seas, signifying the idea of liberty's global application.

There are no roads in the Dutch hamlet of Giethoorn. Canals and food bridges link all of its structures.

55 million years ago, the North Pole had tropical weather complete with alligators and palm palms, according to experts.

Crabs on Australia's Christmas Island are shielded from being smashed by automobile waves during their huge migration by bridges and tunnels.

Males in Egypt are required to serve in the military, but only if they have a brother; if they have no sisters or siblings, they are excused.

Regardless of caste, color, or religion, more than 100,000 people are fed vegetarian meals daily at the Golden Temple in India.

A palava ruler from South India named "Bodhidharma" founded Shaolin Kung Fu in China.

Hollywoodland was inscribed on the renowned Hollywood signage before 1949 when four letters were erased.

The world's largest bridal dress is capable of covering Mount Everest at over 8 kilometers in length.

The bodice maldivica, or palm tree seed, is the biggest seed in the world.

The scariest spot on Earth is not far from San Diego, California. We call it the Rock of Potato Chips.
In a Polish town, every one of the 6,000 inhabitants lives on the same street.

Shigetaka Kurita created emojis in Japan in 1999.

You captured a 72-second radio transmission in space in 1977. We still don't know its origin to this day.

An Australian guy used his typewriter to enter every number between one and one million in words rather than numerals. He finished in sixteen years and seven months.

To produce enough oxygen for every person on Earth for six months would have cost 38 trillion. Trees provide us with this service for free.

The first person to become a billionaire from book sales was J.K. Rowling.

Although villages lack the technological sophistication of cities, they may provide you with mental clarity and rejuvenation.

The most costly spice in the world, saffron, is mostly produced in Iran.

Audi, Bentley, Bugatti, Ducati, Porsche, and Lamborghini are currently owned by Volkswagen.

The location with the longest name on Earth, Taumatawhakatangihangakoauauotamateauripukakapikimaungahoronukupokaiwhenukitanatahu, is situated in New Zealand and has 85 letters in total.

It is so small that it would take two hours to run from West to East and nine days to complete from top to bottom.

Scarlet PCs and cellphones were used by Japan to create all of the medals for the 2020 Tokyo Olympics.

The day Stephen Hawking was born coincided with the deaths of Galileo and Einstein.

Dolphins exhibit bromances when two males form a couple and assist each other in finding female partners for up to 15 years.

An old Greek phrase that means "Goat Song" is where the word "tragedy" originates.

Due to their heightened sensitivity to infrared light, chickens experience daylight around 45 minutes ahead of humans.

China's Hangzhou is home to the world's biggest residential structure. Living in this multi-story complex are over 30,000 individuals.

A sequence of earthquakes in Missouri on February 7, 1812, the most powerful of which caused the Mississippi River to flow backward for many hours.

When a force from the outside breaks an egg, life ends; when a force from inside breaks the egg, life starts.

Norway's public universities are entirely free to attend for any student, wherever in the globe.

The pineberry is a white strawberry with red seeds and a white hue that tastes like a pineapple.

Cats comprehend human directions, according to studies, but they don't always want to comply.

The triangular border between Argentina, Brazil, and Paraguay is the most amazing border on the planet.

Because pineapple includes proteins that break down flesh, if you put a slice of pineapple in your mouth, it will begin to devour you.

Thirteen young men commit themselves as a result of Princess Qaja, a beauty icon in Persia, rejecting them.

Outside of the house where he grew up, the apple tree where Isaac Newton discovered gravity is still standing.

Putting a massive mirror 10 light years distant from Earth and using a telescope to view it supposedly will allow us to see 20 years into the past.

A frog cannot throw up its stomach contents if it consumes anything poisonous. Rather, the frog undergoes a condition known as complete gastric eversion in which it regurgitates its whole stomach.

Mihai Eminescu, onesti, Romania is home to one of the most inventive sculptures in the whole globe.

The youngest person to visit every nation on Earth and achieve a world record is Lexi Alford.

Strasbourg saw a pandemic of dance in 1581. Approximately 400 individuals had heart attacks while dancing wildly.

With a temperature of -93.2°C, Dome Fuji in Antarctica is the coldest spot on Earth.

Iran has a hamlet called Masouleh. where the dwellings lie underneath the streets.

The black rose is an incredibly uncommon flower that grows wild only in the Turkish town of Halfeti.

The actor Chris Hemsworth, who portrayed Thor in the Avengers films, has a strong affinity for India. India is even the name he gave his daughter.

One of the holiest and most significant Hindu sanctuaries in Bali is Pura Besakih, which is used by Hindus around the island.

In Jakarta, Indonesia, there is a spa where clients get massages using pythons and snakes.

In case of an asteroid strike, Oreo has built an asteroid-proof bunker in Norway where they would preserve their recipes and cookies.

China has said in the open that they have created an artificial sun that is 150 million degrees hotter than the sun.

QR codes are often seen on gravestones at cemeteries around Japan. You may see the person's life story on video after scanning the code.

The Krzywy Domek, a crooked home in Poland, is preserved for its distinctive architectural design.

The magpie is one of the most intelligent bird species on the planet; it can even identify itself in a mirror.

3D carpets are used in German hotels to stop guests from racing down the corridor.

With over 19% of the world's wine produced there, Italy is the world's biggest producer.

You may simultaneously touch North America and Europe in a diving trench.

Google Maps finds the greatest natural bridge in the world. The name of the bridge is Xian Ren Qiao.

The band did remain on board the real Titanic and performed music while it sank. Every single one of them passed away.

Horses known as gray Percherons are born black and gradually become gray as they age.

Located in West Java, Indonesia, the Citarum River is regarded as the world's most polluted river.

Even though he won the French Scrabble championship and learned every word in the dictionary, the guy from New Zealand still cannot speak French.

During the Vietnam War, the US unleashed more than 2 million tons of bombs on Laos between 1964 and 1973. For nine years, this comes out to one aircraft load every eight minutes.

The base of the Oregon octopus tree is fifteen meters broad, and its estimated age is three hundred years.

The only nation in the world that has enough space to accommodate 114% of its people in an emergency is Switzerland.

In Ketchikan, Alaska, eighth-grade students do a two-night survival test on an uninhabited island as their final physical science exam.

An unusual occurrence known as "Sun Dogs" occurred in China, when three suns emerged and lingered in the sky for around three hours.

The world's longest tunnel, spanning 57 km, connects Switzerland and Italy under the Alps. The building took seventeen years.

The winner of the 2019 Wildlife Photography Prize is a picture of two mice who look to be fighting in London's subway system.

There have only ever been five gold medal-winning Olympians; four of them have earned nine gold medals. Holds: Michael Phelps, 23.

If you reserve any vacation package with Delta Airlines during the month of your birth, you will get a $100 discount.

The Royal Institute of British Architects has only given a royal Gold medal for architecture to one city worldwide, and that is Barcelona.

Three animals—a sheep, a duck, and a rooster—were the first people to ride in a hot air balloon.

It is referred to as "poisonous" if you bite it and die, but as "venomous" if you bite it and die.

Mansa Musa was the wealthiest man to have ever lived; at the time of his death in 1331, his fortune was $400,000,000,000.

"Everything Men Know About Women" is a book with merely 120 blank pages. The author is Dr. Alan Francis, a renowned psychologist in America.

The only location in the world, regardless of nationality, where you may live and work without a visa is Svalbard, Norway.

When a Sunday occurs on a national holiday in Japan, the next working day is off for the public.

In certain countries in 1960, whistling was considered a compliment for ladies.

You may vent your rage by smashing dishes against the wall at a restaurant named Isdaan Floating Restaurant.

In the eighteenth century, bringing a dog or cat to feed the lions was a way to get free entry to the London Zoo.

More than 6 million people's bones are interred in Paris's catacombs, a subterranean grave.

A genre of music known as Binaural Beats exists. This music has the power to elevate you and induce unique dreams.

The only city in India devoid of money, politics, or religion is Auroville. There, people live in harmony, and the area is well-known as the "city of dawn."

There are roughly 200 deceased on Mount Everest, and their remains serve as beacons for climbers.

Brazil is the most biodiverse country on Earth, home to over 50,000 distinct species of plants and trees.

The Roche limit states that if the Moon were moved 21 times closer to Earth, it would shatter and produce rings around the planet.

Your white blood cells and the white dot you see floating in the sky.

On Earth today, there are still 14 trees that existed before Jesus was born.

The alpha-x train, located in Japan, is the fastest in the world.  It can reach a maximum speed of 360 km/h (225 mph).

Julia Tuttle was the only woman to build a significant American city when she built the City of Miami.

With about 700 rivers and tributaries, Bangladesh has an extensive and complex river network spanning more than 24,000 km.

Only Jamaica has a flag made of black, green, and yellow instead of red, white, and blue.

A river runs through the center of the Villagio retail center in Doha, Qatar, and the surrounding planting is designed to resemble the sky.

Asymmetry spanning 1,500 acres is said to conceal the remains of 5 million people.

Horses have a very lengthy soft palate that functions as a one-way valve, which prevents them from throwing up.

A zorse is a hybrid between a female horse and a male zebra.

"Cargo" refers to the cargo carried by ships, whereas "shipment" refers to the freight transported by automobile.

The fourth funnel of the Titanic was a fake, designed to give the ship a more symmetrical, strong appearance.

Three times higher than Mount Everest, the Olympic Mountains are the highest peak in our solar system.

We will need to include the planet of our birth on our passports when the first child is born on Mars.

Every letter in the English language is used in the statement "The quick brown fox jumps over the lazy dog".

The route passes through the United States Death Valley and is more than 200 kilometers long.

For three weeks, the first astronauts to set foot on the moon were placed in quarantine in case they had brought any dangerous viruses with them.

Hurricane Mitch forced the Choluteca River to be redirected, despite Honduras having constructed a bridge across it in 1998.

We could have generated enough energy to power the whole planet if solar panels had covered even 1.2% of the Sahara desert.

Petrichor is the earthy scent released when rain falls on the ground.

The man who refused to salute Hitler, August Landmesser, was eventually forced into prison military duty, where he was murdered in combat.

Only humans would chop down a tree, make paper out of it, and write "save trees" on it.

The popular Nigerian actress Osita Iheme has made a ton of memes for everyone.

Though he seems childlike due to an uncommon ailment, he is an adult.

In Dubai, the government offers 3 grams of gold for every kilogram of weight lost. "you lose, you gain".

Matt Gone has inked both of his eyes in blue and green, covering 98% of his body in tattoos.

You face the death penalty in North Korea if you are found to be selling or watching pornography.

The longest cake in the world was made by Keralan bakers. It is 27,000 kg in weight, more than 6 km long, and 10 cm broad.

Another moniker for Martin Laurello was the "human owl." He could move his head 180 degrees from birth.

The longest national dinner without any repeats is Switzerland's.

Taking away the extra scenes and credits The duration of Titanic is precisely two hours and forty minutes, which is the exact duration of a real-life ship sinking.

The magic tap is a top that can be found at Aqualand Santa Maria, Spain.

Russia is the only nation on Earth where one region experiences day and another experiences night.

At about 19 inches tall, the "flat mobile" is the lowest street-legal automobile in the whole globe.

Many attractive ladies in the Brazilian village of "Noiva De Cordeiro" need a husband, but there aren't enough males in the area.

Throughout their stay, visitors at Dubai's Burj Al Arab Hotel are given a 24-karat gold iPad.

The size of Central Park in New York is almost twice that of Monaco as a whole.

A new world record was achieved by Pakistani students who planted nearly 50,000 trees in 40 seconds.

The UN claims that cutting someone off from the internet violates their human rights.

Taiwan has constructed a glass church shaped like a high-heeled shoe, rising to a height of 16 meters (55 feet).

In Ecuador, there's a swing perched above a cliff. It's known as the swing at the end of the world as it lacks any safety features.

Saudi Arabia is home to the biggest hotel in the world, the Abraj Kudai, which has 10,000 rooms, 70 restaurants, and five rooftop helipads.

In 13.48 seconds, the world's fastest 7-year-old kid completed a 100-meter sprint.

Because marine limestone forms the peak of Mount Everest, the highest point on Earth was once under the water.

A saguaro cactus may live up to 200 years, and the growth of its first arm takes 70 years.

The first component of self-control is hunger. Everything else is within your control if you have control over what you eat and drink.

Terminator is the name of the line that divides Earth's day and night sides.

Men stare at women for almost a year of their lives, according to a poll.

200 times stronger than steel, yet a million times thinner than paper, "graphene" is the strongest substance in the world.

In China, there is a reverse zoo where people are housed in cages while the animals are free to wander.

The Australian army was involved in the Great Emu War of 1932, which ended in defeat for the humans.

The only object on Earth with the ability to do that much harm is a hydrogen bomb, which is capable of killing millions of people.

To combat air pollution and global warming, China is establishing the first forest city in history, where every structure is covered with a million plants and trees.

The Earth's atmosphere gives the impression that the sun is yellow, but in reality it is white.

More than 20% of the oxygen in the world is produced in the Amazon rainforest.

Before relocating to New York City, Sylvester Stallone trained his daughters in self-defense by hiring naval seals.

At Grand Canyon Caverns, 220 feet below the surface in a 65 million-year-old cave, lies the world's deepest, darkest, oldest, and quietest hotel room.

Because chickens are the only animals that synthesize the protein needed to form egg shells, scientists have now concluded that chickens evolved before eggs.

Since the desire to be free is a fundamental human drive, there is no penalty in Germany for a prisoner who attempts to get out of custody.

A 19-year-old trash guy won $15 million in the lotto in 2002. He is back to being a trash guy after blowing all of it on drugs, gambling, and prostitution.

The 41-year-old Georgian Etibar Elchyev, often referred to as the "magnet man," smashed his previous record of 50 spoons by holding 53 spoons on his chest and back.

The son of the wealthiest man in Asia was abducted by a Hong Kong mobster who demanded a ransom of $130 million. The gangster then called the victim and sought investment advice.

To get lottery winnings, a Jamaican lottery winner donned a spooky mask. so that his family won't harass him for money.

There is a 102-year-old ship in Sydney's west that has been transformed into a floating forest.

The reason why the sticker outlasts the product is because the stickers labeled as "Made in China" are created in South Korea.

The Yosemite Cascade has a yearly burst of sunlight that gives the cascade a firefall appearance.

24-carat gold may be consumed without becoming sick. Pure gold will pass through the body and be excreted as waste since it is chemically inert.

The arid lowlands of Antarctica will be the driest spot on Earth. That area has not seen rain in more than two million years.

Because Iraqis think that readers do not take books and criminals do not read books, book merchants on the streets of Iraq keep books open in their stores at night.

Cricket saw the usage of testicular guards for the first time in 1874, and helmets for the first time in 1974. Men didn't understand the importance of the brain until a century later.

The richest structure, often referred to as the eye of the sharara, is the most unusual location on Earth.

Pilots and copilots aboard airplanes eat separate meals to prevent food poisoning from occurring at the same time.

Jackie Chan's mother was a drug dealer who he met via an arrest, and his father was a spy.

In the history of movies, New York is the city that gets devastated the most.

When being transported, helicopters are shrink-wrapped to protect them from sea spray damage.

At a New York auction, a Mark Rothko artwork named "Yellow and Blue" brought approximately $46.5 million.

Both residents and tourists are stealing bricks from the Great Wall of China to construct residences and keep them as mementos.

The readiness paradox states that although being ready for risk lessens the harm it does, the threat is seen as not being as serious as it first seems.

On a court that floated amid a Persian Gulf lagoon, Rafael Nadal and Roger Federer once engaged in a tennis match.

To guarantee that one brother would live to carry out their flying experiment and to prevent the possibility of a double tragedy, the Wright brothers only flew together on a single trip.

An estimated 600 lightning strikes are reported to the Statue of Liberty annually.

The world's biggest mirror is created on Bolivia's Uyuni Salt Flat during the rainy season by the rainwater.

Those who excel in pubs are also highly educated people outside of the game. Your superiority in the game demonstrates your IQ.

Google Maps uses the speed at which Android smartphones travel the road to determine traffic.

Psychology holds that people are never content, regardless of how wealthy they get or how many of their goals come true.

Because liquid oxygen is magnetic, a strong magnet may be used to move or even pick it up.

The world's most extreme lavatory is situated in Serbia, above sea level, and perched on a rock overlooking 8,500 feet.

Queen Victoria detested being pregnant, found nursing repugnant, and felt newborns were unsightly.

Dinosaur, Colorado is a tiny town in the United States of America. Tyrannosaurus Trail, Stegosaurus Freeway, Tyrannosaurus BLVD, and Tyrannosaurus Bypass are a few of its street names.

Approximately 90% of international commerce is conducted by water. China is the greatest exporter of products in the world, and it owns seven of the ten biggest ports worldwide.

The misspelling of Googol, which is one followed by 100 zeros, is the name Google. This was chosen to indicate that the search engine was meant to provide a lot of information.

The tomato is the fruit consumed worldwide the most. After reading this, you may be under the impression that a tomato is a vegetable, but in actuality, they are both classified as fruits and vegetables.

In addition to keeping the crocodiles' teeth clean, the Egyptian plover bird consumes food that gets lodged between their fangs.

The National Palace Museum in Taiwan is home to the biggest collection of Chinese art and antiques in the world, with over 7,000 items.

The skull of a historical Portuguese serial murderer has been kept in a jar at the University of Lisbon for 181 years in startlingly good condition.

"Merchandise 7x" is the code name for a hidden component found in Coca-Cola. Since John Pemberton created it in 1886, it has been a mystery. The description is vault-locked.

Since they were seen to be technical doping, Nike was prohibited from using a pair of shoes they had developed.

A guy by the name of Sam Panopoulos invented Hawaiian pizza, the first kind of pizza made with pineapple, in Canada in 1962.

Reza Paratesh, an Iranian impostor of Lionel Messi, was charged with using his false identity to have sex with twenty-three women.

Speaking with your mother may assist in lowering stress levels and has the same impact as hugging her. Children hear their moms' voices and emit the hormone oxytocin.

Dina Sanichar was found in Uttar Pradesh in 1872 by a party of hunters. The child was pursuing a group of wolves while on all fours.

The study of fungus, including its genetic and biochemical characteristics, classification, and applications to humans, is known as mycology in biology.

Alexander the Great ruled over all of Asia, Persia, Egypt, and Macedonia concurrently as pharaoh.

The sturdy but flexible maple wood used to construct NBA courts allows players to leap and land safely.

The legendary sprinter Usain Bolt can accelerate faster than the Toyota 601 L. Bolt outpaces automobiles with his rapid speed.

Japan is in the process of creating a unique technology that will enable dwellings to be raised above the ground to provide earthquake protection.

The Mudanjiang City mega farm in China, which spans an amazing 22,500,000 acres and is nearly the size of Portugal, is the biggest in the world.

The world's richest cities are Tokyo, London, and New York, respectively. Tokyo has the highest concentration of millionaires, multimillionaires, and billionaires.

There is no meteor crater in Xico. In Mexico, it was a volcano, then a lake, and once the lake dried up, agriculture replaced it.

Bart Jason, a Dutch artist, builds a drone out of his deceased cat Orville. His aerial endeavors envision people soaring through the skies on their animals.

Every number must be written out in words; the letter B won't be used until you have reached one billion.

There are three vaginas in kangaroos; the main one is used for giving birth, while the side ones transport sperm to the two uteruses.

The blobfish's unsettling look has earned it the title of ugliest animal in the world. They are residents of Tasmania and Southeast Australia.

Since the establishment of the modern Olympic Games in 1896, the United States has won more than 1,000 gold medals overall.

Cats who develop a tuna addiction and become "tuna junkies" won't eat anything else.

China is allegedly experiencing the world's worst brain drain. Ten percent of students who study abroad never return to China to reside.

People who suffer from asthma, bronchial obstructions, and other breathing issues go to subterranean salt mines for relief.

Surgeons in Australia carried out the first heart transplant using a dead heart in 2014.

For around six weeks, Heath Ledger holed himself in a motor room to get ready for his part as the Joker.

Researchers believe that bumblebees might be able to fly higher than Mount Everest and reach the top of Earth's peaks.

You can prevent dry scalp, reduce dandruff, and strengthen your hair by mixing vodka into your shampoo.

In Arizona, you might face up to 25 years in prison for felling a cactus. It's like chopping down a species of protected tree.

At the summit of Africa's Mount Kilimanjaro are enormous groundsels that resemble a scene from the Jurassic Park movie.

It was suggested in a Middle Ages German folklore that kissing a donkey may relieve dental pain.

The age of the universe is estimated by an international team of astrophysicists to be 13.8 billion years old.

Nearly all greyhounds have a blood type that permits all other canines to utilize their blood.

You only need to shower once or twice a week, for health reasons, not to mention appearance or odor.

Africa is where millions of tons of electronic garbage from across the globe are dumped in landfills.

Chemical processes in the brain as well as the muscles and nerves in your head and neck are what create headaches.

The majority of the fastest runners in the world are descended from a single Kenyan tribe known as the "Kalenjins".

People cross bridges that have grown rather than ones that were constructed in the wettest regions on earth.

Africa is greater than the combined areas of China, the USA, India, Mexico, and a sizable portion of Europe.

About 50 to 100 hairs are typically lost by humans each day. People who have thick or long hair may notice hair loss more readily.

The Pacific Ocean is experiencing a trash surge, with an estimated area roughly equivalent to Texas.

To increase his supply of cocoa during World War II, an Italian pastry maker combined chocolate with hazelnuts to create Nutella.

Households pay an annual TV license fee in two-thirds of Europe, half of Asia, and Africa.

Because they are unable to chew, ants cut food by moving their jaws sideways, much like scissors, to remove the meal's fluids.

Black Hawk, a Sauk native American chieftain who fought for the British in the War of 1812, is the inspiration for the name of Black Hawk helicopters.

Canals originated in China. China's Grand Canal was established in 486 BC and spans 1,103 km.

The sticky bandages were created by Earle Dickson in response to his wife's repeated self-cuts while housework.

Six persons were murdered in the Great Fire of London in 1666, which burned 13,200 dwellings, 87 churches, and the Saint Paul Cathedral, leaving 70,000 people homeless.

Among the world's finest theoretical physicists is Sabrina Gonzalez Pasterski. Stephen Hawking, the renowned theorist, has praised her work, earning her the moniker "New Einstein".

On July 10, 1913, the hottest temperature ever recorded on Earth was recorded at Greenland Ranch in Death Valley, California, USA.

Intense running exercise may cause the body to accumulate lactic acid, which can alter the flavor of breast milk.

There are breaks in the Great Wall of China. It is made up of hundreds of kilometers' worth of disjointed walls.

Since the 12th century, the unicorn—Scotland's national animal—has represented power and purity.

Wombat excrement has a cubic form. This peculiar form marks the wombat's territory and aids in the poop's retention.

The Sahara is not the biggest desert on the planet. The biggest desert on the planet is officially Antarctica. There isn't much precipitation there.

Because Japanese consumers love their phones so much that they use them in the shower, 90 to 95 percent of mobile phones sold there are waterproof.

75% of what individuals say while they are intoxicated is true.

When a dragon blood tree is cut, it spills blood similarly to a person.

The ocean generates 70% of the oxygen needed for life on Earth via the photosynthesis of phytoplankton and algal plankton.

By burning fat stores at the base of its tail, an alligator may spend up to two years without eating.

Barbara Soper, a Michigan lady, gave birth on 08/08/08, 09/09/09, and 10/10/10. 15 million to one is the odds.

Wood is more expensive and scarce in the whole universe than diamonds.

Kinkajou is a kind of mammal that can sprint up and down trunks with ease by turning its feet backward.

In Norway, there is a bridge called "The Twist," which is essentially an inside museum.

To reuse the same water for the flush, it is customary in Japan to wash hands on the toilet. In this manner, Japan saves millions of liters of water annually.

Because the company prioritizes its clients and only a limited number of cars are available, staff are not permitted to purchase Ferrari vehicles.

An Irish instructor holds the record for the loudest item ever uttered, having screamed the word "quiet" at 121 dB, which is the same as a jet engine.

Don Ritchie's act of kindness, of giving individuals on the ledge a cup of tea and a talk, prevented almost 500 people from taking their own lives.

According to neurologists, each time you control your anger, you are teaching your brain to be more loving and peaceful.

Panthers are not a true species; rather, they are melanism, the reverse of albinism, in jaguars and leopards.

A survey found that men hide in the toilet seeking peace for an average of seven hours a year.

The renowned Hollywood trilogy Lord of the Rings was shot in Hobbiton, Shire, New Zealand.

Tigers are among the most vindictive creatures on the earth, and they can and will exact retribution on anyone who has harmed them.

The Kansas City Public Library is one of the most stylish libraries ever.

Mike Hayes, a student at the age of 18, raised money for his schooling in 1987 by pleading with 2.8 million individuals for one cent.

Ants evolved more than 100 million years ago, and they have two stomachs instead of ears and lungs.

China constructed the highest air purifier in the world, the 100-meter-tall anti-smog Tower, which can generate around 10 million cubic meters of fresh air every day.

Some individuals possess the ability to deliberately cause their vision to become fuzzy or unfocused.

Sylvester Stallone's Rocky movie turtles are still with him. They are older than forty.

The only site on Earth where one may see the sun setting on the Atlantic and rising on the Pacific is Panama.

Charlie Chaplin's voice was first heard on film when he sang a gibberish song in contemporary times.

One of the biggest flowers in the world, Amorphophallus titanium blooms for about four days once every forty years.

The point on Earth closest to the sun is the summit of Chimborazo.

Russian scientists have gathered and examined 300 ancient worms from the Arctic permafrost. Two of the worms, one dated at 32,000 years old and the other at 41,700 years old, started to move and feed after being thawed.

A bite from a blue-ringed octopus does not have antivenom.

An 800-year-old Norwegian stave chapel dating back to 1181 was constructed completely out of wood and did not use any nails.

With just 30 residents, Hum, the smallest town in the world, is located in Croatia.

The Verzasca River in Switzerland has water so pure you can see to the bottom.

One biscuit made it through the Titanic's demise. It brought $23,000 at auction in 2015.

A gift from France, the Statue of Liberty was dedicated on October 28, 1886, and it stands for democracy and freedom.

Born into slavery, Madam CJ Walker rose from poverty to become America's first female self-made billionaire in the 1910s.

More than 400 billionaires participated in psychological research, which revealed that wealthier individuals do have better lives. Also, they discovered that individuals who work for their riches are happier than those who get it via inheritance.

The cardiac muscle that makes up our hearts is unique to the heart and is never weary.

Your runny nose after weeping is caused by the lacrimal punctum, a tiny hole in your eyelids that allows secretions to escape into your nose.

Despite not having a nose at birth, Eli Thompson is still one of the beautiful babies.

Croatia's Galesnjak is the most ideal heart-shaped island.

The biggest underwater sculpture in the world is located in the Bahamas.

The capital of caves in Europe is Granada, Spain.

Thirty thousand years ago, the Homotherium saber-tooth tiger became extinct.

A snail could safely scuttle down the edge of a razor blade because of the protective slime that snails secrete.

Like 3D Job and Max in London, the longest and biggest piece of 3D street art in the world was produced.

The small strings inside bananas are phloem bundles. They have an important purpose, to distribute nutrients throughout the fruit.

Japan utilizes heated sidewalks to combat heavy snowfall, ensuring safe pedestrian passage and reducing snow removal costs.

Barcode scanners scan the white spaces in between and not the black bars.

There is a wine fountain in Italy that dispenses free wine 24 hours 7 days.

In 1908 the Russians arrived 14 days late to the Olympics because they were using the wrong calendar.

A pomato is a grafted plant that has cherry tomatoes on the vine and white potatoes in the soil.

Sound in ATMs is not produced by rollers delivering the cash. The sound is produced by a speaker to reassure you that money is on its way.

In the Middle Ages, books were so valuable that libraries would chain them to the bookcase. This was widely practiced until the 18th century.

In Texas distance is measured in time not miles.

The last coach of the train is marked with a yellow X paint mark. This mark is the signal for the train supervisor to know that the entire train has Departed and no coaches left behind.

Emeralds are more than 20 times rarer than diamonds and therefore often command a higher price.

Goats have accents and if you take two goats from different parts of the world they won't be able to understand each other.
Bruce Lee's speed made it difficult for cameras to capture his moves so he had to slow down his movement for television.

The Victoria water lily is the strongest leaf in the world and it is capable of holding up a human.

If humans had a vision as good as an eagle, we would be able to see an ant on the ground from the roof of a 10-story building.

We would also have brilliant color vision, UV vision, and nearly a 360 range of view.

In Japan, all the old phone booths are converted into aquariums.

Female anacondas eat male anacondas after mating so that they can survive in the seven months of fasting during pregnancy.

A 17-year-old boy from Mexico City Julio Macias died after his girlfriend gave him a love bite. The suction resulted in a blood clot that traveled to the brain causing a stroke.

In Thailand, it is illegal to go out in public if you are not wearing underwear.

In South Africa, white people own 72% of the country's farmland, even though they are 8% of the population.

Graphene Aerogel is the world's lightest
material it is seven times lighter than air.

There are stones in Romania called servants
that grow and multiply by rain showers.

Iceland has no public railway system.

In Germany, it is illegal to run out of fuel on
the Autobahn. If you are caught you can face
a fine or even a driving ban.

The average person will spend 6 months of
their lifetime waiting for a red traffic light to
turn green.

MSC Irina is the largest continental ship in
the world; its length is 1,312 feet and can
carry up to 24,346 containers.

The average adult will work almost 75,000
miles over their lifetime, the equivalent of
traveling around the world three times.

Most diamonds are at least 3 billion years old. There are enough diamonds in existence today to give everyone on the planet Earth a cupful.

The world's largest ocean, the Pacific Ocean, is so big that it is larger than all of Earth's land area combined.

There are 30 km of tunnels underneath downtown Toronto, connecting shops, subway stations, and restaurants so that people do not need to go outside into the heat and snow.

The Fastest Japanese train, the Yamanashi Maglev, can travel over 1 km in 7 seconds or 1 mile in 11 seconds. It is said to become operational after 2027.

Cats can drink seawater which is salt water. Unlike humans, they have kidneys that can filter out salt and use the water content to hydrate their bodies.

China is the most irreligious country on Earth. A Staggering 90% of residents claim no interest or relevance to religion whatsoever.

The world's first animated feature film was made in Argentina.

The Barbados thread snake is the smallest known snake species.

Ford has made a new perfume which smells like petrol. The perfume is named Ford Mustang Mach-Eau. Mach-Eau is made for electric car users who will miss the smell of petrol.

All birds find shelter during the rain but eagles avoid strain by flying above the clouds.

Nick Vujicic was born without arms or legs but he is a painter, swimmer, skydiver, motivational speaker, and pastor.

There's a tree in South Africa known as the Tree of Life and is over 6,000 years old.

The largest print photograph ever taken was 111 ft (34m) wide and 32 ft (9.8 m) high.

It snowed in the Sahara desert for 30 minutes on the 18th of February 1979.

Humans are bioluminescent and glow in visible light. However, the light that we emit is 1,000 times lower than the sensitivity of our naked eye so we are not able to see it.

A tree named Great Banyan in China is 250 years old 80 ft tall and covers 3.5 acres of land.

An advanced transparent solar panel has been made that can replace house windows and even our phone screens.

The oldest known animal species is the all-shoe crab which has been around for 450 million years.

The Guinness Book of World Records holds the record of being the book most often stolen from public libraries.

The world's smallest jail is in Ontario Canada it is just over 24 meters.
Japan is known as the land of the rising Sun.

A soda bottle plus sunlight plus water plus chlorine equals a 55-watt light bulb.

More than 800 languages are spoken in New York.

The world's most expensive perfume is Clive Christian's No. 1 Passant Guardant. It costs

143,000 for 30ml and comes in a flat studded with 2,000 diamonds.

There's a scenic highway in China that is completed over water.

Bhutan became the first Nation in the world to ban the sale of tobacco and smoking in public places.

Beers don't pee or poop while hibernating. The waste is broken down into proteins used to preserve muscles and tissue during long sleep.
Dubai Miracle Garden is the largest natural flower garden in the world. It has more than 50 million flowers.

There is a Shop in Hunan province in China, 120 m (393 ft) up off the side of a cliff it supplies climbers with essential snacks, refreshments, and sustenance during their ascent.

Apple is so rich that it can buy us
The world's smallest cow Rani, is a 23-month-old dwarf cow in Bangladesh that is 20 inches tall and weighs 28 kg.

The unique smell of rain comes from plant oils, bacteria, and the ozone.

A white rainbow is also known as a fog bow and a ghost rainbow.

The human body has about 10% of hydrogen.

Pumpkin is not a vegetable scientifically it is a berry.

The air conditioning budget for US soldiers in Afghanistan was $20.2 billion. This is more than NASA's entire budget.

Dragonflies are one of the fastest insects flying 50 to 60 mph.

The world's largest grand piano was built by a 15-year-old in New Zealand. The piano is a little over 18 feet long and has 85 keys, three shots of the standard 88.

Toothpaste removes ink marks. Just apply it to stains and let it dry before you wash it.

If you find a bug in the Xbox Live network, Microsoft will reward you up to $20,000.

In the city of Setenil in Spain inhabitants live under the largest rock in the world.

Venustraphobia is a fear of a beautiful woman.

In the 1930s Nikola Tesla reportedly invented the death ray. This Ray was capable of causing mass destruction and killing so to save Earth from this he destroyed his invention and the plans for the laser were never found after Tesla's death.

Singing in the shower helps boost your immunity, lower your blood pressure, reduce stress, and improve your mood.

The British pound is the world's oldest currency still in use. It is 1,200 years old.

It is illegal to run out of gas in Youngstown, Ohio.

In Russia, women outnumber men by 10 million.

A scorpion can hold its breath for up to 6 days.

Three Gorges Dam in China is one of the most massive dams in the world.

The world's most expensive cheese is made from donkey's milk.

In 1859 24 rabbits were released in Australia within 6 years, the population grew to 2 million.

Not only do mosquitoes bite you and suck your blood but they also urinate on your body before flying off.

Jellyfish and lobsters are considered biologically immortal. They don't age and will never die unless they are killed.

If you cut out the brain of a tortoise it can survive another 6 months.

The sketch that Jack drew of Rose wearing the famous necklace in the Titanic movie was drawn by director James Cameron.

Baby elephants don't know how to use their trunks to drink water until they are 9 months old.

Each year Disneyland uses over 5,000 gallons of paint to maintain the clean appearance of the park.

HP printer black ink is more expensive than human blood.

There is a hidden beach in Mexico that was caused by a bomb in World War I.

Human saliva contains a painkiller called opiorphin which is six times more powerful than morphine.

Worthersee Stadium, Austria soccer stadium is filled with a 300-tree Forest to draw attention to environmental issues.

"&" and "AND" mean different things in movie credit. Two writers' names joined with "&" means they collaborated while "AND" means they worked on the script at the same time.

Chloroform takes about 6 to 8 minutes to make a person unconscious and not just 10 seconds as shown in the movies.

You can survive entirely on a diet of potatoes and butter which provide all the necessary nutrients the human body needs.

The rarest blood type in the world is RH-Null blood also known as the golden blood it is only been identified in 43 people during the last 50 years.

About 10% of the world's population is left-handed.

A 2,000-year-old green serpentine mask was found at the base of a pyramid in Mexico.

If you want to separate water and alcohol once they are mixed, put some table salt in the solution you will obtain two different layers.

Almost 14% of Google employees have never attended college.

A supermarket in Thailand uses innovative banana leaves to package food instead of plastic to avoid unnecessary plastic waste.

The VBX6 BMW X6 is painted in the world's blackest black. Vantablack is one of the darkest substances known to absorb over 99% of light.

Nearly half of the Ellen pilots admit to falling asleep during a flight. 29% of them said they woke up and saw the other pilot asleep as well.

In 2015 Beijing banned 2.5 million cars for 2 weeks to get a beautiful blue sky for a World War II parade.

If you heat a magnet, it will lose its magnetism.

The decomposition of cellulose and lignin, two chemical components in paper, gives old books their pleasant fragrance.


**YOU CAN ALSO READ OR BUY OTHER BOOKS BY: JAMES A. HENDERSON THAT ARE AVAILABLE ON AMAZON.**

**1. Title: A Brief History Of Queen Elizabeth II**
**Subtitle: A Life Well Spent.**

**2. Title: Native American Historical Battles.**
**Subtitle: A History From Beginning To End.**

**3. Title: The Ukraine War**
**Subtitle: The Untold Truths And Latest Updates On The Ukraine And Russia War.**

www.ingramcontent.com/pod-product-compliance
Lightning Source LLC
Chambersburg PA
CBHW071924210526
45479CB00002B/552